Animals That Migrate

Animals That Migrate

Carmen Bredeson

Franklin Watts
A Division of Scholastic Inc.
New York • Toronto • London • Auckland • Sydney
Mexico City • New Delhi • Hong Kong
Danbury, Connecticut

For Dr. Donna Shaver, a sea turtle's best friend.

Note to readers: Definitions for words in **bold** can be found in the Glossary at the back of this book.

The photograph on the cover shows wildebeests crossing the Masai Mara River in East Africa. The photograph opposite the title page shows horseshoe crabs in Delaware Bay, NJ.

Library of Congress Cataloging-in-Publication Data

Bredeson, Carmen
    Animals that migrate / Carmen Bredeson
        p.   cm.— (Watts Library)
    Includes bibliographical references and index.
    ISBN 0-531-11865-7 (lib.bdg.)   0-531-16573-6 (pbk.)
    1. Animal migration—Juvenile literature. [1. Animals—Migration.] I. Title. II. Series.
QL754 .B74 2001
591.56'8—dc21

00-051343

Contents

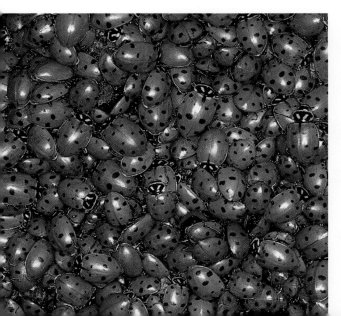

The ruby-throated hummingbird migrates across the Gulf of Mexico each year.

Chapter One

The Mysteries of Migration

Every year gray whales swim from Alaska to Mexico and back, while tiny hummingbirds fly all the way across the Gulf of Mexico twice. Animals that go from one place to another and then back again each year are said to be migrating. Some animals migrate to mate and reproduce. They often go back to the place where they were born to lay eggs or have their young.

Many animals migrate to warmer climates during the winter. It is usually not the cold that sends these animals south. They have feathers or fur that could keep them warm. Instead they go south searching for food. Plants do not grow very well in cold temperatures and many of the insects that animals eat die during the winter.

How do animals find their way across hundreds or thousands of miles of land and sea? Nobody knows for sure, but there are many theories. Some species could use the Sun and stars to guide them from place to place. Others might be able to tune into Earth's magnetic field to point them in the right direction.

Birds seem to follow landmarks such as coastlines, rivers, or mountain ranges in their trips north and south. Smell and taste might also play a role in **navigation**. Maybe animals use all of these methods to help them migrate or maybe they use only one or two.

Scientists who want to study migrating animals use many methods to track animals in the wild. Information bands are placed on the legs of some birds. When the birds are found in another area, the bands tell scientists how far they have flown. Radar is also used to follow the path of large groups of birds. Radar can tell how large a flock is, but it cannot tell what kinds of birds are in the flock.

Some animals that scientists want to follow are fitted with radio collars or **transmitters**. These devices give off signals that are picked up by antennas on Earth. The signals from some types of transmitters can even be picked up by satellites orbiting hundreds of miles above Earth. Tracking data from the satellites is beamed to computers back on Earth, allowing scientists to follow the movements of animals such as whales, sea turtles, and wildebeests.

Even with all of this tracking, scientists still cannot explain exactly how migrating animals find their way on these amazing journeys. But studies of migrating animals can give us information on where animals migrate and why these animals move from place to place.

Red crabs pass by residents of Christmas Island during their yearly migration.

Chapter Two

Fish and Crustaceans

How would you like to wake up during the night and have a crab crawling across your pillow, or even your face? That happens to some of the people who live on Christmas Island. This tiny island in the Indian Ocean is home to more than 100 million red crabs. Once a year the crabs leave their **burrows** in the hills and head for the sea to mate and release eggs. During this annual migration, the 2,500 people who live on Christmas Island better watch out!

*Red crabs on
Christmas Island
migrate to the beach
each year to mate.*

As soon as the rainy season arrives in October or November, the crabs begin their journey. Large males lead the parade, followed by females and younger crabs. A hundred million crabs swarm across roads, through yards, and even into houses. If someone accidentally leaves a door open, the crabs click and clack themselves right into the house. Then residents have to sweep the intruders off the furniture and out from under tables.

When the crabs reach the beach they pile into the ocean to soak up water. Their long journey has left them dried out and weak. Some crabs even use their big claws like cups and dip water into their mouths. After getting a good drink, male red crabs crawl onto the sand and dig burrows, where they will mate with the females.

After mating, the males take another dip in the water and head back to their forest homes. Female red crabs stay in the burrows on the beach for 12 to 14 days, waiting for the eggs in their **brood pouches** to **mature**. When the eggs are ready, the females crawl down to the water after the Sun has set.

Larvae, Larvae Everywhere

Each mother crab shakes her body back and forth to break open up to 100,000 eggs in her brood pouch. She makes a squeaking noise as her **larvae** are released into the water. Millions of tiny crab larvae, which look like little shrimp, cloud the water along the shoreline. Once the larvae are released, the

13

The larvae of red crabs look like tiny shrimp.

mother crabs begin their journey home. Their entire trip from the hills to the beach and back takes over a month.

The larvae that are left behind spend almost a month in the sea. As they mature into crabs, many of the larvae are eaten by fish and birds. When they are about 0.2 inches (0.5 centimeters) long, the little crabs scramble ashore and begin their trip up to the hills. The millions of tiny crabs look like a moving red carpet on the ground. They are even harder to keep out of the houses than their parents. Christmas Island residents find little red crabs in the kitchen cabinets, on lampshades, under the beds, and even swimming around in the toilets!

Finally, to the relief of the islanders, all of the crabs make it back to hills and into the forest around Christmas time. Things settle down for the people on Christmas Island until the next year, when the red crab migration begins all over again.

Where Are All the Eels Going?

American and European eels live quiet lives in freshwater rivers, lakes, and ponds. As autumn approaches each year, some mature eels begin to gain weight and change color from yellow to silver. They are preparing for a long migration that takes them down rivers, into **estuaries**, and eventually into the Atlantic Ocean. In their rush to get to the sea, eels sometimes have to slither across short areas of dry land to get to another body of water. Changing from freshwater to saltwater does not seem to bother the eels.

After arriving at the ocean, the eels from Europe and the eels from America all head for their breeding ground in the Sargasso Sea. This large area of calm water is located between Bermuda and Puerto Rico in the Atlantic Ocean. The Sargasso Sea is named for the large mats of **sargassum** seaweed that float on the surface of the water. For American eels, the trip is about 1,000 miles (1,609 km) long. European eels, though, travel up to 3,000 miles (4,828 km) to the breeding ground.

This American eel in Vortex Springs, Florida will eventually migrate to the Sargasso Sea in the Atlantic Ocean.

During March and April, thousands of adult eels arrive in the Sargasso Sea to **spawn**. After the female eels release their eggs into the water, the males fertilize the eggs. According to University of Georgia zoology professor Gene Helfman, "for females, the bigger they are the more eggs they contain. A large female may have more than a million eggs." After spawning, the adult eels die.

From Elvers to Eels

The fertilized eel eggs hatch into **transparent** larvae that are 0.25 inches (0.64 cm) long. The tiny creatures do not have digestive systems, but seem to absorb food through their skin. The larvae drift with the ocean currents and gradually turn into small eels, which are called elvers. Elvers are about 2.5 inches (6.4 cm) long and are about as thin as a pencil lead.

American elvers reach estuaries along the coast of the United States after about a year of drifting. European elvers

Spiny Lobsters All in a Row

When the first autumn storms hit the waters of the western Atlantic Ocean, thousands of spiny lobsters gather together in the sea. They line up to march into cooler, deeper water where they will be safe from large waves. Instead of heading out in a group, though, the lobsters go single file across the sea floor. Each one uses its antennas to keep in touch with the lobster in front.

Spiny lobsters travel up to ten miles a day and march for several days without stopping. If an enemy approaches and threatens the lobsters, they gather in circles with their tails together and their bodies facing out to scare the predator away.

spend up to 3 years drifting with the currents before reaching Europe. After arriving at the coast, elvers mass together in large groups to begin their long swim upriver. They move into freshwater lakes, rivers, and ponds, where they grow into yellow eels that can reach a length of 6 feet (1.8 meters). Ten to twelve years later, the full-grown eels feel an urge to begin moving toward the ocean. They head for the Sargasso Sea to mate and die, just as generations of their ancestors did before them.

Elvers travel from the Sargasso Sea where they were born to freshwater lakes, rivers, and ponds.

Each year, thousands of Laysan Albatross nest on Midway Island in the Pacific Ocean.

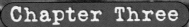

Chapter Three

Birds

In the month of October, thousands of Laysan Albatross arrive on the island of Midway in the Pacific Ocean. These birds gather on this remote island to mate, build their nests, and raise their young. Each female lays just one large egg, which **incubates** for 2 to 3 months. When first hatched, the chicks have to be kept warm by their parents. As soon as the little birds begin growing feathers, they can stay warm by themselves. Then their parents leave for days at a time to search for food at sea.

By the time the chicks are 5 or 6 months old, they are ready to soar out to sea like their parents. Except for the mating and nesting season, the Laysan Albatross lives at sea. Its 7-foot (2-m) wingspan allows it to glide on air currents for hours at a time. The Laysan Albatross can fly and glide for thousands of miles without ever resting on land.

The Laysan Albatross eats fish, squid, and anything else it can find floating on the surface of the water. The birds come back to Midway Island each year to breed and have their young. This pattern had gone on for generations with no problems, until World War II.

During the war, the United States Navy thought Midway Island was the perfect place to build an airfield. They got to work and constructed runways, airplane hangars, and barracks for the servicemen who were stationed there. Dozens of planes took off and landed on the long runways of Midway Island each day. When October arrived, thousands of migrating Laysan Albatross started landing too. They perched on airplane wings and all over the runways.

Laysan Albatross Versus the U.S. Navy

The Laysan Albatross is nicknamed the gooney bird. It is graceful in the air because of its long wingspan, but it is a clumsy bird on land. Because of its large size, the gooney bird has trouble getting airborne sometimes. The birds especially like the long runways of the Midway Island airfield. Running along the smooth surface helps them get into the air.

During World War II, the Navy had a war to fight and the birds on Midway Island were in the way. Navy personnel put up scarecrows to try to chase the birds away, but that didn't work. They tried capturing birds in nets and taking them to other islands. The birds returned. Had the gooney birds defeated the United States Navy? It started to look that way until someone made a discovery. If vehicles were driven down the runway with their horns blowing and sirens blaring, the birds took off for a few minutes. So, during the Laysan Albatross nesting season navy personnel had to scare the birds into the air long enough for each airplane takeoff and landing.

More than half a century after the end of World War II, the Laysan Albatross is still arriving at Midway Island each year.

The Laysan Albatross, or gooney bird, has a wingspan that is 7 feet (2 m) long.

21

During World War II, Laysan Albatross had to be cleared from the runway to allow military airplanes to take off and land.

At least 800,000 of the birds migrate to the island, which is halfway between the United States and Japan. The military base is closed now, and the island is a protected area. There is still one runway that is used for commercial airplanes. And yes, during nesting season the birds have to be cleared from it before a plane can take off or land.

The Arctic Tern Flies Pole to Pole

Amazingly, Arctic terns spend eight months a year in the air! The birds live in the Arctic during the northern summer when the Sun never sets. As soon as the weather begins to turn colder and the days get shorter, the terns head south on an 11,000-mile (17,702-km) journey to the Antarctic. When they arrive four months later, it is summer at the South Pole and daylight for 24 hours a day.

Danger in the Sky

Many birds die during their annual migrations. Bad weather and strong headwinds can leave the birds so exhausted they fall from the sky. Some who land on the ground recover, but those who fall into the water often drown. Cities also create dangers for birds. Many bird species fly at night, and may use the stars to navigate. City lights confuse the birds and sometimes cause them to crash into the sides of buildings or other tall structures.

Arctic terns spend 8 months of the year in flight between the Arctic and Antarctic.

In late February, with the approach of winter in the south, the birds head north again. They arrive in the Arctic in early June, during the summer. The Arctic terns spend only about four months of the year on solid ground. The rest of the time they are in the air, making them the long-distance champions among migrating birds.

Ruby-Throated Hummingbirds

The ruby-throated hummingbird weighs only one-eighth of an ounce, yet it flies non-stop across 500 miles (805 km) of the Gulf of Mexico during its annual migration. When winter approaches in North America, the birds head south to Mexico and Central America. The hummingbird's wings beat more than 50 times a second during the 18 to 20 hour journey. Because the wings beat so fast, a hummingbird is able to make some pretty amazing moves. It can hover in mid-air, zip forward, stop suddenly, and fly backward. A hummingbird heart beats up to 1,200 times a minute, pumping blood to its wing muscles. The tiny birds are exhausted and hungry when they arrive at their winter homes.

Flowers are what hummingbirds look for, especially red flowers. The birds' pointed beaks can poke deep into flowers and the fuzzy tips on their tongues are great for lapping up nectar. A hummingbird eats 50 percent of its body weight in nectar each day. In addition to flowers, the birds like to dine on small insects and spiders.

After spending the winter eating in Mexico and Central

America, the ruby-throated hummingbird returns to North America to lay eggs in the spring. Females build tiny nests of moss, tree bark, fuzzy fibers from plants, and spider-web silk. They usually lay only two pea-size eggs, which hatch into featherless, blind, hungry chicks. The mother bird is kept very busy collecting nectar and small insects for her brood. Male hummingbirds do not help gather food for the young.

Chicks start growing feathers in about 6 days and their eyes open a couple of days later. By the time they are four weeks old, the chicks are ready to fly out of the nest. They are not yet on their own, though. Their mother follows them for several more weeks, feeding them until they grow larger and learn to feed themselves. By winter, the young ruby-throated hummingbirds will be strong enough to make the long flight over the Gulf of Mexico to their winter feeding grounds.

The ruby-throated hummingbird uses its pointed beak to eat nectar from flowers.

Kemp's ridley sea turtles migrate to the beach to lay eggs. When the eggs hatch, the baby turtles move across the sand toward the water.

Chapter Four

Reptiles and Amphibians

The Kemp's ridley sea turtles are the smallest and most endangered sea turtles in the world. The turtles have one main nesting ground, a strip of beach in Rancho Nuevo, Mexico. Over the years, fewer and fewer female ridley turtles migrated to this beach to lay eggs. Biologists were afraid that a hurricane or an oil spill might destroy the beach and the nesting grounds and wipe out the species completely.

This female Kemp's ridley sea turtle is laying eggs at Rancho Nuevo, Mexico.

In order to try to save the ridleys from extinction, an interesting program was started in 1978. People at the Rancho Nuevo beach began watching for female turtles coming ashore to nest. When a turtle was spotted, her eggs were collected before they ever hit the bottom of the pit she had dug in the sand. The gathering of the eggs did not bother the female turtles since they go into a kind of trance while laying about one hundred eggs, each the size of a Ping-Pong ball.

A Trip to Padre Island

After the eggs were collected, they were carefully carried back to camp and put into a box of sand from Padre Island. This barrier island is located along the lower Texas coast. The boxes of eggs were flown to Padre Island and kept in a hatching barn, where they were carefully watched.

About 50 days later, silver-dollar-size turtles emerged from the eggs. The little black hatchlings were carried to the beach and put down in the sand. With flippers waving, the tiny turtles stumbled and fell all over each other in their rush to get to the water. After a brief swim the turtles were scooped up in little nets, put into boxes, and taken to Galveston, Texas.

The ridleys spent the next 9 months living in a turtle barn at Galveston, growing and getting stronger while they swam around in buckets of water. When they were the size of a Frisbee, it was time for them to be released in the Gulf of Mexico. Just before their release, each turtle had a numbered silver tag attached to one of its flippers. If the turtle was ever found in the wild, the tag would identify it.

What was the reason for this complicated program? Turtles usually return to the place where they were born to mate and lay eggs. Scientists hoped the ridleys that were hatched at Padre Island and allowed to crawl to the water would remember the area and return there to lay eggs. Then there would be a second nesting ground in case the one in Mexico was destroyed. About 13,000 Kemp's ridley turtles were hatched at the Padre Island location from 1978 to 1988.

It takes turtles 10 to 15 years to mature. During those years, the people at Padre Island waited to see if their experiment was working. In 1996, the first two tagged ridleys were found laying eggs on the beach during the May to July nesting season. That number increased to four tagged turtles in 1999. In the summer of 2000 only one tagged turtle was found. A lot of seaweed covered the beaches that year, though, making turtle tracks hard to spot. Hopefully more ridleys will migrate to Padre Island to lay eggs in the future. With the help of their human friends, the species might survive after all.

Dens Full of Red-Sided Garter Snakes

The Canadian province of Manitoba is home to thousands of red-sided garter snakes. These non-venomous snakes live farther north than any other North American species of snake. When the weather turns cold, thousands of these snakes crawl into underground dens in the rocks. They twist themselves together into huge wiggling masses before settling down to a winter of **hibernation**. As many as 20,000 snakes may gather in one den. For about eight months of the year they stay underground without eating or drinking.

As the ground warms during the months of April or May, the red-sided garter snakes begin to stir. The males emerge

These red-sided garter snakes are emerging from hibernation.

from the den first. They slither out in large groups and gather around the entrance to the den, waiting for the females. Female red-sided garter snakes crawl out one or two at a time. They are immediately surrounded by as many 100 males who are waiting to mate. After mating, the snakes crawl off alone to their summer feeding grounds in the prairies and marshes.

Snake Feast

The red-sided garter snakes fill up on fish, frogs, rodents, worms, and **tadpoles**. They need to eat enough during the short summer season to last them through next winter's hibernation. Some of the snakes travel as far as 11 miles (18 km) to feed. That is pretty far for a snake that is only about 20 inches (51 cm) long.

Young red-sided garter snakes are born at the summer feeding grounds during the month of August. Female red-sided garter snakes do not lay eggs. Instead they bear live young. A female red-sided garter snake bears an average of eighteen young, but can bear as many as eighty. After the young snakes are born they must take care of themselves because the mother snake abandons her young right after birth. The little snakes eat and grow during the rest of the summer. When September arrives, the red-sided garter snakes return to their dens and get ready to sleep the winter away.

Thousands of "Quacking" Frogs

When the weather warms in early spring, quacking fills the air in some areas of Canada, Alaska, and the northern United States. But those are not ducks quacking, they're wood frogs! Wood frogs come in many shades of tan and brown. Because they have a dark mask that surrounds their eyes, they are sometimes called "robber" frogs. They are only 2 inches (5 cm) long when full-grown and would easily fit into a coffee cup.

After hibernating all winter under piles of leaves and dead tree branches, wood frogs are ready to mate. The males leave

Wood frogs leave their homes in the woods when they are ready to mate.

their homes in the woods and hop to nearby ponds and puddles. They surround the pond and begin their strange "quacking" calls as they wait for the females. When the females arrive at the ponds, they are surrounded by male frogs, who are eager to mate with them. After mating, each female spends about a week laying a jelly-like mass of nearly one thousand eggs in the water of the pond. When their work is finished, male and female wood frogs return to the woods. They do not stay in the pond to care for their young.

From Tadpoles to Frogs

Eggs left behind in the water take 2 to 3 months to hatch into small black tadpoles. The little tadpoles eat some of the jelly that surrounds the egg masses. After emerging from the jelly in May and June, the tadpoles eat algae and plants as they begin to turn into frogs. They start to grow little legs and their tails are absorbed into their bodies. Once tiny frogs are formed, they are ready to begin life on land. Thousands of little frogs hop out of the ponds and make their way into the woods. There they eat spiders, bugs, snails, and slugs as they grow into adult wood frogs.

Frog Tunnels

When frogs begin their annual migration in the spring, some have to cross roads to get to their breeding ponds. Many of these frogs are killed by cars and trucks. In order to make the frog migration easier, people have built tunnels under the roads in some places. The frogs can safely hop through the tunnels instead of trying to cross busy roads and highways.

Manatees, which are also called sea cows, are gentle mammals that eat plants and move very slowly through the water.

Chapter Five

Mammals

Manatees are slow, gentle animals that are sometimes called sea cows. They cruise through the water at a slow rate of speed and must come up to breathe every 5 to 10 minutes. They sleep with their heads down and tails at the surface of the water. The average adult is 10 to 13 feet (3 to 4 meters) long and weighs 1,200 to 3,500 pounds (544 to 1,588 kilograms). In spite of their large size, manatees are not at all aggressive. The West Indian manatee lives in the warm waters off the southeast coast of the United States and in the Gulf of Mexico.

This manatee mother and baby are eating plants in Crystal River, Florida.

Manatees spend their lives floating in the water and slowly munching on sea plants. They can eat their way through as much as 15 percent of their body weight a day! All of that chewing wears out their teeth. "So the manatee has hit on the trick of having throwaway teeth. Throughout the animal's lifetime, new teeth keep erupting at the back of the jaw and move forward as the worn-down teeth in front fall out," says anatomy professor Daryl Domning.

Danger in the Water!

As soon as sea water temperatures begin to fall in the winter, manatees slowly make their way into **lagoons** and estuaries close to shore where the water is warmer. A manatee cannot survive in water much colder than 56 degrees Fahrenheit (13 degrees Celsius). They are able to live in either salt or fresh water and easily migrate between the two.

Manatees communicate with each other using high squeaks and chirps. They seem to have trouble hearing

Follow the Bubbles

A lot of gas is produced in the manatees' intestines because they eat so many plants. You can sometimes find a manatee in the water by following a trail of gas bubbles.

low-frequency sounds like those made by boat motors. Many manatees are injured or killed by boats, especially when they swim close to shore. Manatees that manage to avoid boats sometimes live to be 50 to 60 years old. Females usually have only one calf at a time, after a 13-month **gestation** period. The calf stays close to its mother for at least a year while she migrates into warm water in the winter and back out to open water in the summer.

An Ocean of Wildebeests

It is September in East Africa and more than one and a half million wildebeests are on the move. They stir up huge clouds of dust as they make their way across the dry Serengeti Plain. Deep grooves are worn into the soil as millions of hooves pass by. The animals are following the smell of rain. There is tender green grass poking up through the soil where rain has fallen. The long dry season has left the wildebeests thin and weak and they quickly gobble up the young plants.

During their journey south, herds of wildebeests must cross the mighty Mara River. Many animals are swept away and drowned when they plunge into the raging river. Others are killed by predators such as alligators and lions that wait for the migrating herds to arrive each year. Hundreds of thousands of wildebeests cross the river safely, though, and continue on their trek.

After reaching the grasslands of the southern Serengeti in November, the wildebeests stay in the area for about

Fifty-Pound Baby!

Wildebeests usually have only one calf at a time, but it can weigh nearly 50 pounds (22.7 kilograms) at birth.

This migrating herd of wildebeests is crossing the Mara River in Kenya.

A wildebeest mother eats grass on the Serengeti Plain with her 1-day-old calf close by.

6 months. During this time nearly 500,000 calves are born. Within a few minutes after birth the wobbly calf can stand on its own. By the time it is fifteen minutes old, the calf can run.

Eventually the rains stop in the south and the wildebeest herds must move again to find food. They head westward across the Serengeti toward Lake Victoria and stay in that area until all of the grass is gone. Then it is time to move north again, back across the Mara River. It is July by now and the beginning of the dry season. The herd travels north, eventually arriving at the place where they started. The entire trek has lasted nearly a year and taken the wildebeests across more than 1,800 miles (2,897 km). Their journey will start all over again, as soon as the smell of rain fills the air in September.

Voyage of the Gray Whale

It is late September and the water in the Bering Sea off the West Coast of Alaska is starting to freeze. That is the signal for more than twenty thousand gray whales to begin their 6,000-mile (9,656-km) journey south. It will take the whales about three months to reach their winter home in the warm waters near the Baja peninsula of Mexico. The gray whale travels farther than any other mammal during its annual migration. Because the whales swim fairly close to shore, many people gather along the West Coast of the United States to watch their journey. Adult gray whales weigh as much as 40 tons and can grow to be 50 feet (15 m) long.

After reaching the warm waters of Baja in late December, pregnant female whales give birth to their young, which are called calves. It has taken more than a year for the babies to develop inside their mothers' bodies. Female whales have only one calf at a time, usually every 2 to 3 years. At birth, baby gray whales are about 15 feet (5 m) long and weigh 1,000 to

*Gray whales can weigh
as much as 45 tons.*

1,500 pounds (454 to 680 kg)! The young whales grow quickly, gaining 50 to 100 pounds (23 to 45 kg) a day.

The adults need very little food during this time. Before they left Alaska, they had eaten several tons of tiny shrimp-like creatures, called krill, from the ocean floor. This large amount of food was stored in their bodies as fat, or blubber. The gray whales survive on this stored food during their migration.

When spring arrives in Baja, the whales know to begin their annual trip back to the Bering Sea and their main food supply. The baby whales are strong enough by then to make the 80 to 90 day journey. They swim along beside their mothers, often sticking their heads up out of the water to look around. The mothers stay close to the coast to protect their young from the killer whales that live in deeper water. The gray whales arrive at the Bering Sea in late May. They begin eating tons of krill to put on fat for their next journey, which will begin again in September.

These ladybugs are clustered together in the Santa Catalina Mountains in Arizona.

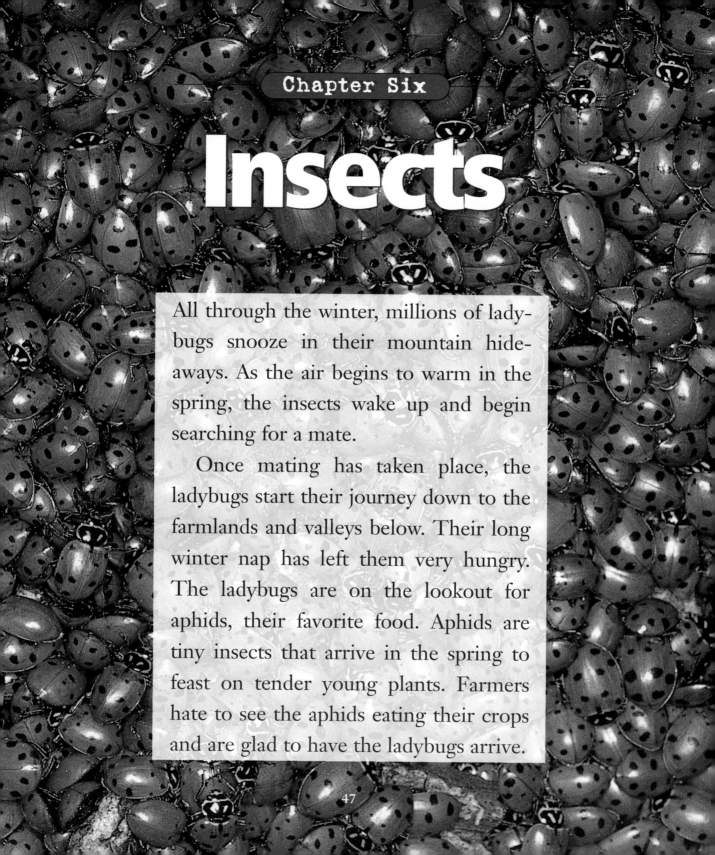

Chapter Six

Insects

All through the winter, millions of ladybugs snooze in their mountain hideaways. As the air begins to warm in the spring, the insects wake up and begin searching for a mate.

Once mating has taken place, the ladybugs start their journey down to the farmlands and valleys below. Their long winter nap has left them very hungry. The ladybugs are on the lookout for aphids, their favorite food. Aphids are tiny insects that arrive in the spring to feast on tender young plants. Farmers hate to see the aphids eating their crops and are glad to have the ladybugs arrive.

Gourmet Aphids

While nibbling on delicious aphids and other insects, female ladybugs lay groups of ten to fifty yellow eggs on the undersides of leaves. A few days later the eggs hatch into blue-gray colored larvae. It takes these wingless larvae about a month to change into adult ladybugs. During that time the larvae gobble up aphids too. For the rest of the summer, young and old ladybugs spend their time eating and sleeping among plants in the fields.

When the weather begins to turn cooler in the fall, clouds of ladybugs take flight from the lowlands. They fly straight up into the hills and mountains where they eat pollen and build

Ladybugs eat pollen before they go into hibernation.

up fat reserves. Then the ladybugs gather together in sheltered places such as caves or tree trunks to begin their winter sleep. As many as one million ladybugs may be clustered together in one place. They hibernate until the warm spring air tells them it is time to mate and head out for their aphid feast.

March of the Horseshoe Crab

It may seem strange that horseshoe crabs are grouped with insects instead of with red crabs and spiny lobsters. Even though horseshoe crabs look like crabs, they are actually distant cousins of spiders and scorpions. These strange, tank-like creatures have been on Earth for at least three hundred million years. Long before there were dinosaurs, there were horseshoe crabs. We know this by studying horseshoe crab **fossils**, the hardened remains of the creatures that lived long ago.

The Atlantic species of the horseshoe crab lives in the ocean off the East Coast of North America. During April or May, thousands of crabs begin to swim toward Delaware Bay, their favorite nesting ground. Females wait for a high tide before crawling onto the beach.

As the female horseshoe crabs lay thousands of eggs in the wet sand, males fertilize the eggs. Females mix sand with the eggs to cover and protect them. This process is repeated several times until the female has deposited as many as 80,000

Horseshoe crabs migrate to the beach to lay eggs.

eggs. Once mating is finished, the adult horseshoe crabs return to the sea. Thousands of clumps of tiny green eggs are left behind on the beach.

Bird Alert!

With perfect timing, hundreds of thousands of migrating shore birds arrive at Delaware Bay on their annual trip north. They are very hungry after spending days in flight. The millions of horseshoe crab eggs in the sand are just what they want to eat. The birds still have hundreds of miles to fly, so they need to build up their body fat. Many birds double their body weight during the week or two they spend eating on the beach. Without the annual horseshoe crab egg feast, many birds would not survive the trip north.

Even though the birds eat millions of eggs, there are still hundreds of clumps left. Horseshoe crab **embryos** develop inside the eggs for several weeks before hatching into little larvae. High tides roll across the beach, carrying the larvae out to sea. Once in the ocean, many larvae are eaten by fish and other sea creatures.

In spite of being dinner for so many animals, some of the larvae survive and develop into horseshoe crabs. For the next 9 to 11 years, the horseshoe crabs remain in the sea, gradually growing into adults. Once they have reached maturity, it is their turn to migrate to the beaches of Delaware Bay to lay their own eggs.

Dinosaur Trivia

Dinosaurs first appeared on Earth about 220 million years ago. They disappeared from the fossil record around 65 million years ago.

This army ant column is crossing a road in Arusha National Park in Tanzania.

An Army of Ants

Army ants do not follow a normal pattern of migration. Instead of returning to the place they migrated from each year, army ants constantly travel to new places in search of food. Columns of as many as a half a million ants move through the night, eating everything in their path. The line of ants can be fifty feet across. They gobble up insects, injured and dead animals, rodents, and lizards. Nothing escapes the march of the army ants. When day breaks, the ants gather together into a ball to rest.

After marching each night for two weeks, the ants set up camp and stay in one spot for nearly three weeks. The queen lays 100,000 to 300,000 eggs in a 5 to 10 day period. Eggs the queen laid at the previous camp have turned into larvae by this time. The larvae have spun **cocoons**, and are going through a process of changing into ants. When the young ants come out of their cocoons they are very hungry. That is the signal for the colony to begin marching again in search of food. Adult ants carry the newly laid eggs as they move out across the countryside. When they come to a body of water, some of the

Army ants carry the growing larvae underneath their bodies as they move from place to place.

53

ants lock their legs together to make a bridge for the other ants to crawl across.

Will the Mystery be Solved?

Wouldn't it be fun to know how animals travel from one place to another without getting lost? Scientists might learn the answers one day. You could even be the scientist who makes the big discovery. Maybe you will be able to tell the world what guides a sea turtle to the beach where it was born or what signals birds follow as they fly overhead. Somebody has to solve the mysteries of migration. Why not you?

Glossary

brood pouches—a cavity in the body of an animal that holds eggs or embryos for part of their development

burrows—a tunnel or hole that an animal digs in the ground

cocoons—a silky case some insect larvae spin around themselves

embryo—an animal in the earliest stage of development before birth

estuary—an inlet or small bay that extends from the sea

fossil—the hardened remains of a plant or animal that lived in the past

gestation—the amount of time it takes for young to develop before birth

hibernation—spending the winter in an inactive state

incubate—to sit on eggs and keep them warm before hatching

lagoon—a shallow pond or lake that is connected to a larger body of water

larvae—the young form of an animal that looks greatly different from the adult

mature—full-grown

navigation—finding the way from one place to another

sargassum—brown algae or seaweed that floats in tropical waters

spawn—to produce groups of eggs or young

tadpole—larvae of some animals such as frogs that have a tail and gills and live in water

transmitters—an instrument that sends information to a receiver

transparent—something you can see through

To Find Out More

Books

Halton, Cheryl. *Those Amazing Eels.* Minneapolis: Dillon Press, 1990.

Lindblad, Lisa. *The Serengeti Migration.* New York: Hyperion, 1994.

Price-Groff, Claire. *The Manatee.* San Diego: Lucent Books, 1999.

Rauzon, Mark. *Hummingbirds.* New York: Franklin Watts, 1997.

Reader's Digest. *The Wildlife Year.* London: Reader's Digest Association Limited, 1991.

Organizations and Online Sites

Crystal River National Wildlife Refuge
1502 SE Kings Bay Drive
Crystal River, FL
See manatees in the protected inlets of Kings Bay.

Field Museum of Natural History
1400 South Lake Shore Drive
Chicago, IL 60605

National Museum of Natural History
Smithsonian Institution
Washington, D.C. 20560

National Wildlife Federation
http://www.nwf.org/wildalive/seaturtle/wwwlinks.html
This web site has information about Kemp's ridley sea turtles.
Click on links to find out more about these turtles on similar
sites.

Padre Island National Seashore
Park Superintendent
9405 South Padre Island Drive
Corpus Christi, TX 78418
Padre Island is a barrier island that is 113 miles long. It is
located on the lower Texas Coast.

Seaworld
http://www.seaworld.org.
Search the Animal Information Database to find out about manatees and other marine animals.

Smithsonian Migratory Bird Center
http://natzoo.si.edu/smbc/
Click on "Featured Birds" to see a variety of colorful bird photographs. Click on the names below the photos to find information about each migrating bird and their habits.

Wake Forest University
http://www.wfu.edu/albatross/index.htm
Check out different albatross studies in Hawaii and the Galapagos Islands. Learn how to get information about albatross from satellites.

Wildlife Conservation Society
http://wcs.org/
Read about animals from all over the world.

A Note on Sources

About the time I started writing this book, the Houston Museum of Natural Science began showing an IMAX movie called *Animal Migration*. Right up on the screen were gray whales swimming along the California coast and giant herds of wildebeests on their annual trek across the Serengeti Plain in Africa. My favorites, though, were the red crabs of Christmas Island. The screen was alive with piles of squirming, clicking crabs making their mad dash to the sea! The pictures were so wonderful I decided the book had to start with their amazing migration.

Also, a few years ago, I got interested in the Kemp's ridley sea turtles and spent some time studying the reptiles. On several occasions I talked to Dr. Donna Shaver, who directs the Padre Island effort to save the ridleys from extinction. Thank you, Dr. Shaver, for answering all of my questions about the work you do for this endangered species.

—*Carmen Bredeson*

Index

Numbers in *italics* indicate illustrations.

About the Author

The amazing things that animals do have always interested Carmen Bredeson. How do tiny birds find their way from the North Pole to the South Pole? And why do so many animals go back to the place where they were born to have their young? Writing *Animals That Migrate* helped the author learn more about the mysterious behavior of animals.

Bredeson is the author of thirty non-fiction books for young people. Her recent Franklin Watts books include *The Moon*, *Tide Pools*, and *Pluto*. She is a former high school English teacher who has a master's degree in library.

The author is pictured on the deck of a Coast Guard ship during a Kemp's ridley sea turtle release in the Gulf of Mexico.